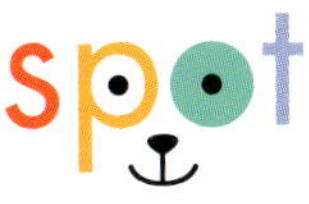

spot

BIG BIRDS

AF575625

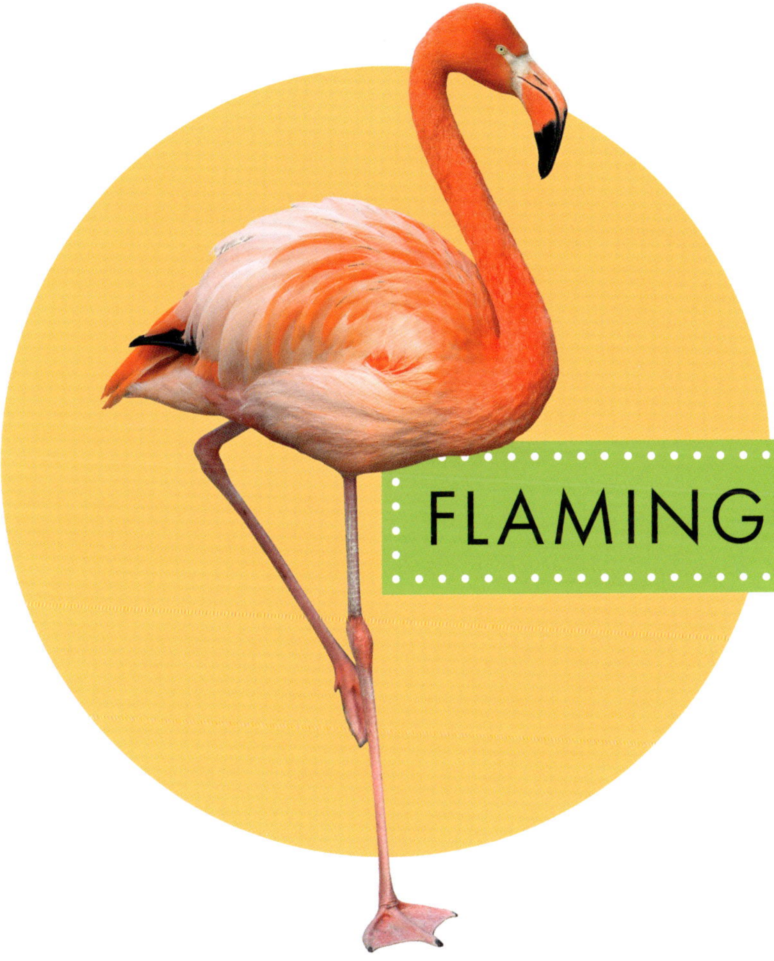

FLAMINGOS

by Lisa Amstutz

AMICUS | AMICUS INK

feathers

bill

Look for these words and pictures as you read.

legs

webbed toes

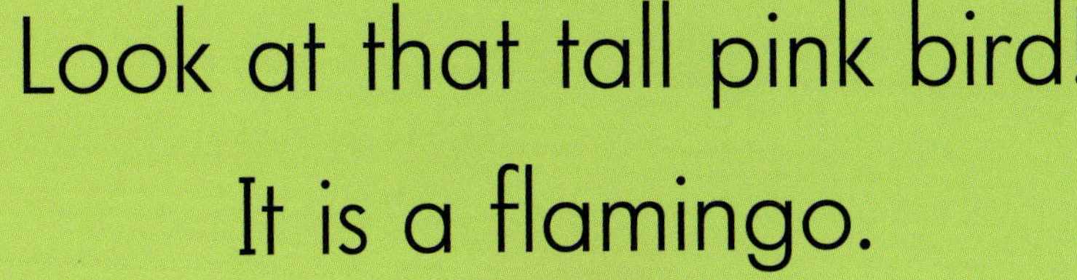

Look at that tall pink bird!

It is a flamingo.

See the feathers?
They help the bird fly.
The bird's food makes them pink.

feathers

The bird dips its bill.
It sucks in water.
Its bill traps plants and shrimp.

bill

legs

Look at the long legs.
The bird holds one next to its body.
The leg stays warm.

Look at the webbed toes. They do not sink in mud.
webbed toes

See the mud nests?
Each nest holds one big egg.

A gray chick peeps.
Mom and Dad give it food.
Soon it will turn pink, too.

See the feathers?
They help the bird fly.
The bird's food makes them pink.

feathers

feathers

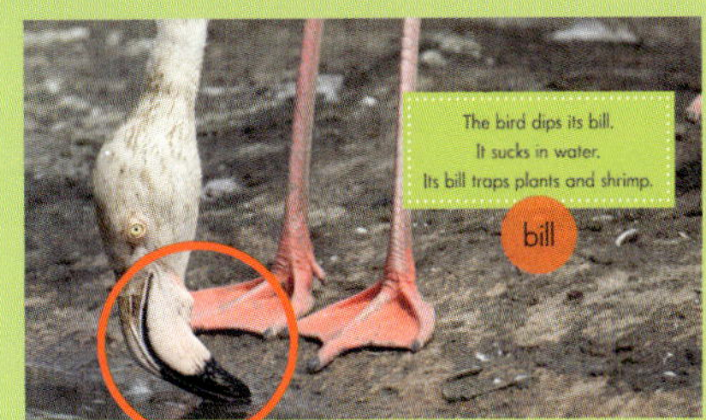

bill

Did you find?

legs

webbed toes

Spot is published by Amicus and Amicus Ink
P.O. Box 227, Mankato, MN 56002
www.amicuspublishing.us

Library of Congress Cataloging-in-Publication Data
Names: Amstutz, Lisa J., author.
Title: Flamingos / by Lisa Amstutz.
Description: Mankato, MN : Amicus, [2023] | Series: Big birds | Audience: Ages 4-7 | Audience: Grades K-1 | Summary: "Pink-feathered flamingos dazzle in this leveled reader about the big bird's habitat, body parts, and behavior. Includes a search-and-find feature to reinforce key vocabulary"-- Provided by publisher.
Identifiers: LCCN 2019060281 (print) | LCCN 2019060282 (ebook) | ISBN 9781645490975 (library binding) | ISBN 9781681526645 (paperback) | ISBN 9781645491392 (pdf)
Subjects: LCSH: Flamingos--Juvenile literature.
Classification: LCC QL696.C56 A47 2023 (print) | LCC QL696.C56 (ebook) | DDC 598.3/5--dc23
LC record available at https://lccn.loc.gov/2019060281
LC ebook record available at https://lccn.loc.gov/2019060282

Gillia Olson, editor
Deb Miner, series designer
Ciara Beitlich, book designer
Bridget Prehn, photo researcher

Photos by Shutterstock / cyo bo cover, 1, 16; iStock / ajansen 3; Dreamstime / Luca Nichetti 4–5; Alamy / Mark Boulton 6–7; Pixabay / Albert Dezetter 8–9; Shutterstock / Diegomezr 10–11; Shutterstock / Sergey Uryadnikov 12–13; Shutterstock / Stephaniellen 14–15